Elisabeth Regina Alves C. Silva

Hydrological potential of the Pontal stream - PE

Elisabeth Regina Alves C. Silva

Hydrological potential of the Pontal stream - PE

Subsidies for the theme of climate change and water management

This book is a translation from the original published under ISBN 978-613-9-64145-1.

Publisher:
Sciencia Scripts
is a trademark of
Dodo Books Indian Ocean Ltd. and OmniScriptum S.R.L publishing group

120 High Road, East Finchley, London, N2 9ED, United Kingdom
Str. Armeneasca 28/1, office 1, Chisinau MD-2012, Republic of Moldova, Europe
Printed at: see last page
ISBN: 978-620-7-72773-5

ACKNOWLEDGEMENTS

I thank God, because without him I wouldn't be here and in him I place my faith, my trust and my love.

I would like to thank my mum and grandma for all the unconditional love they have shown me over the years.

To my supervisor Prof[l] . Dr Josiclêda Domiciano Galvíncio, all my respect, affection and admiration for her availability and patience in guiding me, for the opportunities in my academic life, as well as for the learning I have acquired over the years, which has enabled me to carry out this research. Thank you very much.

To my co-supervisor Prof Dr Hernani Loebler, for his learning and availability during my master's degree. Thank you very much.

I would like to thank the people who directly helped me with my work, such as Rodrigo de Queiroga Miranda, Ygor Cristiano and Pedro Paulo Lima.

I would like to thank my friends for their friendship. Especially my friend José Gustavo, who always helped me during my degree, often doing classwork in my place.

To the SERGEO Remote Sensing and Geoprocessing Laboratory, for providing the fundamental programmes and licences to carry out this work, and to its members. To the National Council for Scientific and Technological Development and to the Postgraduate Programme in Geography at the Federal University of Pernambuco, for the opportunity to develop the project and for the financial support, which contributed to my professional and personal growth.

"When I was five my mum always told me that happiness was the key to life. When I went to school, they asked me what I wanted to be when I grew up. I wrote down **'happy'. They told me I didn't understand the** question, and I told them they didn't understand life" (John Lennon).

SUMMARY

Irrigated agriculture in Brazil's semi-arid region is the main economic activity in the region, which is why in recent decades there has been increased competition for the use of water available for irrigation in this area, with the aim of serving other sectors of society. At the same time, there has been an increase in public policies aimed at increasing the supply of water in the region, through direct investment or public-private partnerships. The big challenge in these areas has not only been to produce, but also: how, when, how much, to whom and where to commercialise. With this in mind, this study sought to make a diagnosis of the Riacho do Pontal-PE Basin, in terms of the natural availability of water for irrigation, and to assess the increase in hydro-agricultural potential that the increase in water to be made available through integration with the São Francisco River Basin is likely to bring. Data obtained from satellite images was used to analyse evapotranspiration in the area over time, showing the growth of the irrigated perimeter in the Basin, and climatic and hydrological data was used to assess hydrological stress and quantify the volume of water to be used in the irrigated areas and in each crop to be implemented in the area after the integration of the Basin, with the aim of optimising available water resources and proposing alternatives for more rational water management and increased agricultural productivity in the Riacho do Pontal Basin.

Keywords: Hydrological stress, fruit growing, optimisation of water resources, agricultural productivity.

SUMMARY

CHAPTER 1

INTRODUCTION

Food security increasingly depends on food production from irrigated agriculture, which makes it irrevocably dependent on water security. According to the United Nations Food and Agriculture Organisation (FAO), 80% of the products needed to meet the needs of the world's population over the next 25 years will be provided by irrigated crops. In this sense, one of the challenges for irrigated agriculture in Brazil is to guarantee water allocations compatible with the demands of soils potentially suitable for irrigation in order to make water supply security compatible with the country's vocation as a food provider, domestic food security and external demand, as well as reducing water losses in irrigation systems, whether in its conduction and distribution in the water infrastructure, or in the application of water to crops through methods and plot management.

Thus, the existence of effective instruments for managing water resources with a view to irrigated agriculture initially presents itself as a problem to be solved. However, the proper application of these instruments can go from signalling regions with development potential for irrigated agriculture to increasing the security of water availability for this activity (MINISTÉRIO DA INTEGRAÇÃO NACIONAL, 2011). According to EMBRAPA (2004), the changes that can be induced by the introduction of irrigated agriculture have implications for the local and regional economic base, depending on the scale of implementation. These implications can have both positive and negative aspects, depending on the social, economic and environmental conditions of the area and the development characteristics of certain projects.

According to projections drawn up by the International Water Management Institute (IWMI), although Brazil is one of the countries with the highest average annual water availability per inhabitant, it falls into the category of countries with economic water scarcity, as it has enough water to meet its needs, However, it has a semi-arid region with poor spatial and temporal distribution of rainfall, which implies the need for investment in the construction of reservoirs for water storage and regularisation and in the construction of conduction systems, as a way of guaranteeing the use of water for sustainable development (MINISTÉRIO DA INTEGRAÇÃO NACIONAL, 2008).

In this sense, knowledge of the ratio between water supply and demand is essential as a subsidy for proposing alternatives for the best use of water resources, given that less than 1.0 million hectares of the 88 million hectares that make up the semi-arid region are suitable for irrigation, considering only local water resources (around 1.1% of the total area). This irrigation potential could be increased by around 2.4 million hectares, rising to 2.7 per cent, if river basins were transposed, which would still mean a very small portion of the semi-arid region (DUARTE 2002; LEITE et. al. 2004).

Identifying the volume of irrigation water makes it possible to assess the efficiency of its use and is an essential element in the sustainable management of water resources. Various studies, such as those carried out by Acreman et al. (2004); the National Water Resources Plan - PNRH (2006); Galvão (2008); Silva and Galvíncio (2011), use hydrological models such as the Macro Water Sharing Plans (MWSP) to help assess hydrological stress, as well as estimating the degree of climate vulnerability and the relationship between supply and demand of water for irrigation in river basins.

Therefore, the main objective of this work is to assess the hydro-agricultural potential of the area corresponding to the Riacho do Pontal Hydrographic Basin in Pernambuco, in an attempt to find better alternatives for optimising the available water resources, with a view to more rational water management. This study will be important in helping to increase agricultural productivity in the region without having to increase irrigated areas to the detriment of the reduction in areas corresponding to caatinga vegetation, thus helping to mitigate the negative environmental impacts resulting from inadequate land use and misuse of water resources.

To fulfil the main objective, the following specific objectives were set:

> Diagnosis of the relationship between water supply and demand in the Pontal Basin - PE through the use of water resources ratio.
> Estimating hydrological stress in the Basin using the Australian MWSP methodology.

CHAPTER 2

LITERATURE REVIEW

2.1 WATERSHED AS A LANDSCAPE UNIT

The watershed can be defined as a physical unit, characterised as a volume (the watershed is an area defined by its dividers, and also has a three-dimensional distribution that begins with land uses, passes through the soil profile and encompasses the rocks that support the watershed) drained by a particular watercourse and limited peripherally by the so-called watershed. Its hydrological role is to transform an inflow of water, concentrated in volume over time (precipitation), into a single outflow of water (runoff). The study of river basins makes it possible to integrate the factors that condition the quality and availability of water resources with their real physical and anthropic conditioning factors (VALENTE, 1976; HEIN, 2000; GROSSI, 2003; PRADO, 2005).

According to Moldan & Cerny (1994), from a hydrological point of view, the microwatershed can be considered the smallest landscape unit capable of integrating all the components related to water quality and availability, such as the atmosphere, natural vegetation, cultivated plants, soils, underlying rocks, bodies of water and the surrounding landscape. Environmentally, it can be said that the watershed is the ecosystem and morphological unit that best reflects the impacts of anthropogenic interference, such as land occupation with agricultural activities (JENKINS et. al., 1994). Lima (1999) comments that the microwatershed is the well-defined manifestation of an open natural system, which can be seen as the ecosystem unit of the landscape.

It is in the river basin that most of the consequences or impacts of the use of natural resources manifest themselves. This is one of the reasons why the basin is taken as the physiographic planning unit. Given the legal provisions inherent to irrigation, water resource management and the environment, any irrigation and drainage project, whether collective (community of irrigators) or individual, must refer to the river basin in which it is located, at least in order to comply with the conditions for granting and, where appropriate, charging for the use of water resources (MINISTÉRIO DA INTEGRAÇÃO NACIONAL, 2008).

2.2 THE PROBLEM OF THE USE OF WATER RESOURCES IN THE WORLD.

Water is a fundamental input for life and an irreplaceable element in various human activities, as well as maintaining the balance of the environment. The accelerated population growth in the world has led to an increase in the demand for water, which has caused water scarcity problems in various regions. It is estimated that currently more than 1 billion people live in conditions where there is insufficient water available for consumption and that in 25 years' time around 5.5 billion people will be living in areas with moderate or serious water shortages (SETTI, et. al., 2001).

The world's water problem has generally been assessed on the basis of the statistics of the Malthusian model (1798), according to which the world's population would be limited in its growth before the end of the 19th century due to a lack of food in the Third World. However, Robert Thomas Malthus did not foresee that thanks to the Green Revolution, the great development of biotechnology and the progressive fall in birth rates, especially in Third World countries, this perspective would become obsolete (REBOUÇAS, 2002).

2.2.1 Basin Integration

One of the solutions to the problem of water scarcity for agriculture has been the integration of basins since the last century. As well as the integration of the São Francisco River in Brazil, other countries have also adopted similar actions to tackle the lack of water supply. There are countless experiences of transposing water from basins around the world, such as the Tejo-Segura in Spain, completed in 1978 after a 40-year completion period (1933 to 1973), with an estimated cost of US$5.3 billion at current values. It transferred water from the Tagus River basin, located on the Atlantic Ocean side of the Iberian Peninsula, to the Segura River basin, a dry region in south-eastern Spain (ARAGÃO, 2008; MINISTÉRIO DA INTEGRAÇÃO NACIONAL, 2014).

The project is 286 kilometres long and has an average flow of 33 m^3 /s. Presented as an example of successful transposition for irrigation and urban supply in the EIA-RIMA prepared by Jaakko Poyry-Tahal, critics claim that the project failed to achieve its main objective and induced an even greater demand for water, requiring new transposition projects to be built and the resolution of problems linked to soil salinisation (ARAGÃO, 2008; MINISTÉRIO DA INTEGRAÇÃO NACIONAL, 2014).

Also an example of transposition was the Snowy Mountains Hydroelectric System in Australia, which has a set of 16 reservoirs, seven power stations, a pumping station and 145 kilometres of tunnels and 80 km of water mains that collect and store water that would normally flow from the east to the coast, being diverted from the Snowy River to the Murray and Murrumbidgee rivers. It was to be completed in 25 years (1949 to 1974) and cost an estimated US$820 million (ARAGÃO, 2008).

The project began in 1949 and includes 16 dams, seven hydroelectric stations, 145 kilometres of tunnels and 80 kilometres of aqueducts. The initial cost of the project, intended for power generation and irrigation, was budgeted at US$630 million. On the other hand, there have been conflicts between the donor and recipient regions, and the transposition currently requires new solutions for water supply, such as wells, reuse and desanilisation (ARAGÃO, 2008).

There are other projects in various countries that are also worth highlighting, such as: The Colorado-Big Thompson Project in the USA, which has a set of 12 reservoirs, 56 kilometres of tunnels and 153 km of canals that transpose the waters of the Colorado River west of the Rocky Mountains to its eastern slope towards the Big Thompson River and which had a completion time of 21 years (1938 to 1959) and an estimated cost of US$1.4 billion; The Lesotho Mountains Hydroelectric Project, on the border between Lesotho and South Africa, with a set of four hydroelectric dams, pipelines and tunnels, with a completion time of approximately 19 years (1983 to 2002) and a cost estimate of US$ 4 billion (the original project envisaged 4 hydroelectric dams and a total budget of US$ 8 billion) (MINISTÉRIO DA INTEGRAÇÃO NACIONAL, 2014).

Also the Wanjiazhai Water Transfer Project: A set of pipelines in the north-western region of Shanxi Province, with three distinct axes measuring 44 kilometres, 100 kilometres and 167 kilometres, drawing water from the Yellow-Huang He River with a completion deadline of 10 years (2001 to 2011) and an estimated cost of US$ 1,5 billion and the Chavimochic Special Project in Peru, with tunnels, open canals, buried pipelines and siphons bringing water to higher regions of the rivers located near the northern coast of Peru, with a completion deadline of 10 years (1986 to 1996) and an estimated cost of $2.15 billion.

There are still projects in the study or implementation phase, such as the El-Salaam canal project in Egypt, which will involve the construction of a 150-kilometre pipeline to transport water from the sewage system mixed with water from the River Nile from the river delta to Sinai, with a completion deadline yet to be stipulated in the project and an estimated cost of US$2.8 billion. In addition to the best known transposition project, the Aral Sea in Central Asia between Uzbekistan and Kazakhstan, which was once the fourth largest inland sea on Earth with 66,000 square kilometres, will have to undergo new interventions (MINISTÉRIO DA INTEGRAÇÃO NACIONAL, 2014).

The diversion of water from the Amu Daria and Sir Daria rivers for cotton plantation irrigation projects, starting in 1939 by the government of the now defunct Soviet Union, consumed 90 per cent of the water that reached the Aral, reducing it to a third of its original size. What used to be a seabed became a desert, with serious impacts on the region's economy, especially fishing. One of the alternatives being studied to restore the Aral Sea is the construction of two canals, one from the Volga River, 800 kilometres long and estimated at US$ 8 billion, and the other from the Ob and Irtysh rivers, 2,500 kilometres long and estimated at US$ 22 million and projected to be completed in 20 years, with an estimated cost of US$ 30 billion (MINISTÉRIO DA INTEGRAÇÃO NACIONAL, 2014).

In Brazil, the big challenge for several years now has been the transposition of the São Francisco River. The Project for the Integration of the São Francisco River with the Hydrographic Basins of the Northern Northeast is a Federal Government undertaking, under the responsibility of the Ministry of National Integration and is intended to ensure a water supply, by 2025, for around 12 million inhabitants of 390 municipalities in the Agreste and Sertão of the states of Pernambuco, Ceará, Paraíba and Rio Grande do Norte and is budgeted at R$8.2 billion. According to the Ministry of National Integration (2011), the system will operate on the two axes of the transposition.

The 402-kilometre (km) North Shaft will take water from Cabrobó (PE) to the Salgado and Jaguaribe rivers in Ceará; Piranhas-Açu in Paraíba and Rio Grande do Norte; and Apodi, also in Rio Grande do Norte. The surplus volumes will be stored in strategic reservoirs in the receiving basins: Chapéu and Entre Montes (PE); Engenheiro Ávidos and São Gonçalo (PB); Atalho and Castanhão (CE); Armando Ribeiro Gonçalves, Santa Cruz and Pau dos Ferros (RN).

In the Eastern Hub, the water will travel a distance of 220 kilometres. It will be captured in the lake of the Itaparica Dam (municipality of Floresta, Pernambuco), and will be carried to the Paraíba River (PB), reaching the existing reservoirs in the receiving basins: Poço da Cruz, in Pernambuco, and Epitácio Pessoa (Boqueirão), in Paraíba. Part of the flow will be transferred beforehand to the Pajeú and Moxotó river basins and to the agreste region of Pernambuco, through the existence of a 70-kilometre branch that will connect the East Hub to the Ipojuca river basin. The maximum expected flow is 28 m^3 /s, but the average operational flow will be 10 m3 /s. The surplus water will be transferred to the Poço da Cruz (PE) and Epitácio Pessoa (Boqueirão, PB) reservoirs.

However, Rebouças (2004) points out that there are several ways of using water correctly and that the abundance of water in Brazil does not authorise waste, much less neglect, because the problem is not just about getting more water, but using what you have wisely. And so, if the traditional idea persists that the only solution to the problems of local and occasional water shortages is to increase the supply of water by building extraordinary works, the water crisis in Brazil could reach unprecedented proportions in the coming years. It is therefore necessary to study ways of increasing productivity with less and less water.

2.3 IRRIGATION PROBLEMS IN BRAZIL

2.3.1 General Aspects

According to the Ministry of National Integration (2008), irrigated agriculture accounts for around % of the total water consumed in most parts of the world. The irrigated area in the Americas is 48,384,878 ha, of which 57.7% is in the United States, 13.3% in Mexico and 6.5% in Brazil. Although the majority of the world's food production takes place in the rainfed system, it has become increasingly necessary to produce food using irrigation, which already accounts for 70% of the fresh water consumed on the planet (FERERES & SORIANO, 2007).

Data from the Ministry of National Integration shows that North America already uses 12% of its water resources for irrigation, while South America only uses 1%. The irrigated area in the Americas is 48,384,878 ha, of which 57.7% is in the United States, 13.3% in Mexico and 6.5% in Brazil. Irrigated agriculture is by far the biggest user: around % of total consumption is attributed to irrigation.

In the United States, 71 per cent of water resources are used for this activity, while in Mexico it is 64 per cent. Although Brazil has approximately 15% of the planet's fresh water, most of this resource (70%) is in the Amazon basin, where only 7% of the Brazilian population lives. More than half of the water consumed in Brazil is destined for irrigated agriculture, although the country's irrigated cultivated area is around 5%.

According to the National Water Resources Plan (2006), 69% of the water consumed in Brazil is used in irrigated agriculture, despite the fact that the country only has around 5% of its cultivated area under irrigation, with an average efficiency of 64%. In other words, 36% of the water used for irrigation in the country is lost through conduction and distribution in hydraulic infrastructures, causing a great deal of waste in the use of water in agriculture. The goal is to produce more food using less water, a task that could be very difficult if climate change projections for the coming decades are confirmed (MARENGO et al., 2009). In agriculture, the FAO estimates that around 60 per cent of the water supplied to irrigation projects around the world is lost through evaporation or percolation (REBOUÇAS, 2003).

2.3.2 History of irrigation in Brazil

Compared to other countries on the American continent, irrigation was a late starter in Brazil. The end of the 19th century and the beginning of the 20th century were marked by the creation of a set of institutions focussed on climate issues, water availability, sanitation and works against bad weather, The first irrigation project began indirectly in 1881, in Rio Grande do Sul, on a private initiative, with the construction of the Cadro reservoir, to supply water to be used in irrigated rice cultivation, which actually began operating in 1903 (BRASIL, 2008).

From the second half of the 1970s onwards, several public irrigation projects were launched in various semi-arid states, benefiting the Northeast region with technological advances provided by hydrological models, and including it in the various stages of development of water resource management (MINISTÉRIO DA INTEGRAÇÃO NACIONAL, 2008). The implementation of irrigated agriculture has produced a new reality in the São Francisco sub-medium. A hostile environment, characterised by irregular rainfall and prolonged droughts, has been transformed into a hub for the production of crops with high commercial value. Until the 1970s, the landscape of the semi-arid region of the Northeast consisted of subsistence agriculture, with a predominance of caatinga vegetation (MINISTÉRIO DO DESENVOLVIMENTO SOCIAL E COMBATE À FOME, 2010).

Until the mid-1980s, fruit-growing in the north-east was involved in agro-industrial development in other areas (the coast) with the production of juices such as cashew, orange, guava, acerola etc. But at the end of the decade, it was realised that fruit prices were more competitive on the international market than the prices of other products on the export list (commodities), leading the government to invest more in fruit-growing for export. With the opening up of trade at the start of the 1990s, the Petrolina/Juazeiro cluster went through unfavourable economic times triggered by the stabilisation plans, as did other segments throughout the country (LACERDA E LACERDA, 2004).

In 1996, the Ministry of Agriculture set up the Programme to Support the Development of Irrigated Fruit Growing in the Northeast - PADFIN (BRASIL, 1997) and later the Fruit Growing Development Programme - PROFRUTA, which aims to raise the quality and competitiveness standards of Brazilian fruit growing to the level of excellence required by the international market (BRASIL, 2002).

Nowadays, Brazil stands out as one of the world's main fruit producers, with a production that exceeds 34 million tonnes; however, our exports are still considered insignificant, given that the country exports only 1.3% of its production, showing an incipient participation in this trade, with 0.3% of the total of US$ 36.8 billion that represents the world fresh fruit market and the rest is sold on the domestic market, denoting that fruit represents an important complement in the diet of the Brazilian population (FERRAZ, 2001; VIEGAS et al, 2004).

Increasing productivity through the use of irrigation has made regional development possible by stimulating the spread of modern production techniques. Initially, through the Northeast Irrigation Programme (PIN), the municipalities of Petrolina/PE, Juazeiro/BA and surrounding areas were selected, mainly due to favourable natural conditions, such as their proximity to the São Francisco River. Two structural models were used for agricultural exploitation: public projects for small family producers and private projects for agricultural companies. The creation of the Irrigation Projects was the result of a government decision to prioritise the internalisation of development, by fixing people in the countryside (MINISTÉRIO DO DESENVOLVIMENTO SOCIAL E COMBATE À FOME, 2010).

2.3.3 *The Federal Government's "MAIS IRRIGAÇÃO" programme*

The experience of most of the public perimeters set up in Brazil has shown that producing is not the most difficult factor to overcome, but rather: how, when, how much, to whom and where to market. According to the Ministry of National Integration (2005), in the projects where business activity coexisted with traditional irrigated agriculture by small producers, the latter were successful, albeit slowly, largely due to the presence of the former as a factor in management leadership. On the other hand, each of the existing public irrigation projects that were built without the issue of the market being a determining factor in assessing their viability, consequently, most of them did not present conditions for sustainability.

Within this context is the federal government's Mais Irrigação (More Irrigation) programme, which is expected to invest R$10 billion, of which R$3 billion will come from public funds and R$7 billion from private initiative. Of the 66 irrigation perimeters planned within the programme, which together total 538,000 hectares spread across 16 states, 32 are under the responsibility of the public company Companhia de Desenvolvimento dos Vales do São Francisco e do Parnaíba (CODEVASF), which is linked to the Ministry of National Integration and has played an important role in the expansion of irrigated agriculture in the country. With the programme, the federal government aims to improve agricultural occupation and infrastructure management in the country's irrigated perimeters.

One of Mais Irrigação's strategic objectives is to maximise the occupation and increase the productivity of irrigated areas; to promote the efficient and sustainable use of water; to make water tariffs more affordable, and to establish partnerships with the private sector, always emphasising support for family farming and small irrigators (CODEVASF, 2007a).

The programme will be important for increasing the area of irrigated agriculture in the Northeast and is divided into four axes: Axis 1, which brings a new exploration model together with public authorities and private initiative, encompasses 8 projects and 189,000 hectares in the states of Bahia, Pernambuco, Ceará, Piauí and Minas Gerais. Axis 2 provides for the implementation and revitalisation of 13 projects totalling around 133,000 hectares spread across eight states (Roraima, Tocantins, Goiás, Piauí, Ceará, Bahia, Minas Gerais and Rio Grande do Sul) (CODEVASF, 2007a).

The public investment planned for this axis is almost R$1 billion, and five of these projects are the responsibility of CODEVASF. The others are the responsibility of DNOCS and the National Irrigation Secretariat (SENIR). Axis 3 includes projects for family farming and small irrigators. There are 27 projects, 25 in the Northeast and 11 under the responsibility of CODEVASF, totalling 61,000 hectares. The public investment planned for this axis is also around R$1bn. These projects are spread across the states of Mato Grosso do Sul, Mato Grosso, Sergipe, Alagoas, Piauí, Bahia, Pernambuco, Ceará, Rio Grande do Norte, Paraíba and Maranhão. Axis 4 brings together 18 projects, nine under Codevasf's responsibility, totalling 155,000 hectares and expected to receive R$89 million in public investment for the studies and projects phase (CODEVASF, 2007a).

2.3.4 *MORE IRRIGATION" programme for Pernambuco*

The prospect is that the irrigated area in Pernambuco will increase from 120,000 hectares to 160,000 over the next two years, contributing to the creation of new agricultural frontiers in the state. Among the upgrading and implementation areas, around 89,800 hectares should receive investment in the state, with a potential investment of R$1.7 billion. Among the projects included in Axis 1 is Pontal. Started in 1996, the Pontal Project covers an area of 33,500 hectares, of which 7,717 will be irrigated. In addition to Pontal, the list of those benefiting from "Mais Irrigação" includes the irrigated perimeters Nilo Coelho (Petrolina), Bebedouro (Petrolina), Boa Vista (between Salgueiro and Terra Nova), Moxotó (between Ibimirim and Inajá), Serra Negra, Terra Nova and the Sertão Canal (stages

14

of the northern axis of the Transposition of the São Francisco River) (CEASA, 2012).

2.4 IRRIGATION PROBLEMS IN THE PONTAL-PE CREEK BASIN

2.4.1 Historical Aspects

There has long been concern about the water supply in the Pontal Basin. In 1962, **the "Agronomic Study and Hydro-agricultural Utilisation Scheme for the Pontal Zone" commissioned by the CVSF** was **carried out, covering an** area of 5,700 ha, located between the BR-122 motorway and the left bank of the São Francisco river. Subsequently, the Federal Government, through DNOS - the National Department of Sanitation Works, considering the perennialisation of rivers in the semi-arid region as an old desire of the northeast, promoted studies on the transposition of water from the São Francisco River to the region, which would involve the states of Pernambuco, Ceará, Paraíba and Rio Grande do Norte. In these studies, the Pontal stream would be perennialised by a derivation from the main adductor system in the municipality of Terra Nova (PE) (CODEVASF, 2007b).

Of the studies carried out, the one that received the most emphasis was the one carried out by the Pernambuco state government, through the CPRH, which even started work on the project to supply 44 m³ /s of water from Sobradinho Lake to perennialise various rivers in the Pernambuco hinterland, such as Pontal, Garças, São Pedro, Brígida and Terra Nova, using the channel of the Pontal stream, a tributary of the left bank of the São Francisco river, as the main conduit. Works such as the Pontal Mother Dam and some sections of the Adductor Canal were started, but have been paralysed since 1982 (CODEVASF, 2007b).

The São Francisco Valley Development Company (CODEVASF), created to replace SUVALE, which had succeeded CVSF, undertook more intensive studies in the Pontal area, carrying out a **"Detailed Soil Survey" in** 1989. The survey covered around 97,000 ha, discontinuously **subdivided into "patches", identifying 54,000 ha of soils suitable for irrigation. This was followed by** pre-feasibility studies for irrigated agriculture in an area of 48,000 ha, subdivided into three sectors: Sobradinho (37,600 ha), Pontal Sul (5,200 ha) and Pontal Norte (5,200 ha). At the conclusion of these studies (1991), it was found to be more appropriate to deepen the design envisaged for the Pontal - South Area and Pontal - North Area sectors, leaving the Sobradinho sector for a later stage, as it presented unattractive economic/financial results (CODEVASF, 2007b).

In 1992, the irrigation areas were defined as 4,000 ha for the Pontal - South Area and 4,200 ha for the Pontal - North Area, based on a catchment in the São Francisco River expected to have a flow of 5.40 m³ /s, confirming the estimates from the feasibility phase. The work was completed at the end of 1993. At the beginning of 1996, work began on the Executive Project for the Pontal Sul Project, which involved optimising the system conceived in the basic project and detailing its construction to a level sufficient for its implementation. However, its optimisation was not limited to this detailing, but extended to the reformulation of various concepts, criteria and characteristics of the basic project, adjusting its conception to the reality of the new projects developed and under development by CODEVASF in the Petrolina region (CODEVASF, 2007b).

2.4.2 Current and future scenario of the Pontal-PE Basin

The Pontal Creek Basin in the state of Pernambuco has been the target of projects of interest to both public authorities and the private sector, due to the prospect of growth in its irrigated perimeter as a result of the integration works with the São Francisco River Basin. In this sense, the most important project for increasing the irrigated perimeter of the basin in question is the Pontal project. This project is located in the valley region of the São Francisco sub-medium, in the area of influence of the Petrolina (PE)-Juazeiro (BA) hydro-agricultural hub, the most important centre for the production and export of irrigated tropical fruit in Brazil. The region's agricultural development is based on climatic conditions, characterised by high sunshine throughout the year, and soils with good suitability for irrigation, which help to promote the quality of irrigated fruit production, meeting not only the demands of the domestic market, but also the requirements of consumers in Europe and North America, where fruit exported from the region is destined (LACERDA E LACERDA, 2004).

Integration into this already consolidated hydro-agricultural environment, with the addition of new production areas, is one of the main justifications for implementing irrigation projects for Pontal Norte and Pontal Sul. The Pontal project is one of the actions of the federal government's Mais Irrigação (More Irrigation) Programme, receiving financial contributions of more than 160 million from the Growth Acceleration Programme (PAC 2). It is located in the rural area of the municipality of Petrolina, in Pernambuco, on the left bank of the São Francisco river. The project covers an area of 27,517 hectares, of which 7,717 hectares are irrigable, 3,588 ha in Pontal Sul and 4,129 ha in Pontal Norte. The Legal Reserve (5,539 hectares) has already been defined and implemented, and the remaining area **is made up of "dryland" land (non-irrigable), canal areas, roads and common use areas.**

The Brazilian government, through its agency CODEVASF, aims to transfer the Pontal areas to the private sector through a Public-Private Partnership (PPP) for the development of the region with intensive irrigated agriculture, investing approximately US$70 million for the partial construction of irrigation infrastructure from the São Francisco River to the Pontal area.

The Public-Private Partnership (PPP) in irrigation will provide agricultural exploitation by guaranteeing the winning bidder the assignment of the real right to use the land, allowing it to be used and charging a competitive irrigation tariff for a period of 45 years. The work on the current integration project is intended to focus primarily on large agricultural companies so that they, in turn, make room for small farmers. The land will be transferred to the winning bidder at no cost, in return for which the concessionaire will have two main responsibilities: to occupy the land within 6 years of signing the contract and to guarantee that the agricultural company will allocate at least 25% of the irrigable land to small farmers who will be integrated into its production chain (CODEVASF, 2013).

When the project is completed, it is expected to generate 7,811 direct jobs and 15,622 indirect jobs. Irrigated agriculture has the potential to grow the following crops: pineapple, cotton, bananas, mangoes, grapes, carrots, beans, beetroot, lemons, corn, watermelons, chillies, milk production, fish and vegetables. Dryland production is also economically viable, capable of generating income and improving the living conditions of the local population. Possible activities include sheep and goat farming (meat, skins, live animals and manure), grain production (maize and beans), cassava production (flour), extractivism (firewood, charcoal, umbu) and lowland crops (such as sweet potatoes) (CODEVASF, 2007a).

2.5 MATHEMATICAL MODELLING

From the perspective of Quantitative Geography, it is necessary to construct models to be used in analysing geographical systems; these theoretically constructed models must be verified and validated with field data using statistical techniques (CHORLEY AND HAGGETT, 1967). A mathematical model has been **defined as a "gathering of concepts in the form of a mathematical equation, which portrays the** knowledge of **natural phenomena" (CONNOLY, 1998 apud PRADO, 2005).**

Regarding the mathematisation of Geographical Science advocated by **Quantitative Geography, Santos (2002) says that, "the use of statistical techniques, if correctly** used, allows for greater precision, since the practical and methodological problems of geography are of such a nature that the use of statistical techniques is suitable for **exerting a strong attraction"**. In this context, the study of the spatial distribution patterns of phenomena (specific events, areas and networks) now forms the basis for quantitative studies of space in which remote sensing and digital image processing technologies have greatly facilitated representations of landscape architecture.

Studies using hydrological modelling, in addition to the use of statistics, are supported by Quantitative Geography (LEITE E ROSA, 2005). According to Tucci (1987), a simulation model can be defined as a representation of the behaviour of a structure, scheme or procedure, real or abstract, which in a given time interval interacts with an input, cause or stimulus of energy or information, and an output, effect or response of energy or information. The aim of modelling is therefore to better understand the processes that occur in a system such as a river basin (FOHRER et al., 2001).

The use of hydrological mathematical models is based on three fundamental conditions: (i) the objective of the study, (ii) the historical data available and (iii) the proposed methodology. The objective of the study defines the level of precision desired for representing the phenomena occurring in the river basin. On the other hand, this precision depends on the quantity and quality of the data available to assess the methodology, so the hydrological model is chosen according to the objective of the study, which will define the desired level of precision (TUCCI, 1987 apud PRADO, 2005).

2.5.1 Hydrological modelling using Macro Water Sharing Plans (MWSP)

Conflicts between the various users of water resources are clearly on the rise in Brazil today. Large-scale examples can be seen in the São Francisco River basin, where water demand projections for irrigation, navigation, the transposition project, human and animal supply and the maintenance of current hydroelectric projects are worrying the availability of water in the river (SETTI et al., 2001).

In addition to concerns about the availability of water in the Basin, there is also concern about future changes in its natural dynamics, caused, for example, by variations in climate patterns. In this regard, the 2007 IPCC (Intergovernmental Panel on Climate Change) report states that the increase in temperature observed since the middle of the 20th century is the result of increased concentrations of greenhouse gases in the atmosphere, caused by human activities. The report estimates a probability of more than 90 per cent that anthropogenic action on nature is the main cause of climate variations, which have become increasingly intense in recent years and are now known as climate change. For the IPCC, the term climate change refers to any change in the climate over a considerable period of time, regardless of whether it is a natural variation or the result of human activities.

According to Souza Filho (2003), climate change has a direct influence on the water resources system and these alterations in the climate end up provoking the need to adopt management practices and instruments for these resources that adjust and adapt to the new climatic conditions. With a view to measuring the susceptibility of river basins to a probable change in climate patterns and the need to know the various ways in which water resources are being used, various methods can be used for this purpose, including hydrological methods (GALVÃO, 2008).

In this sense, understanding the dynamics of basins is crucial for managing water resources, maximising the beneficial results for the population by using hydrological methods to identify problems in the area. Hydrological methods are characterised by establishing restriction flows using only data from historical flow series, on the understanding that this flow is sufficient to maintain certain characteristics of the ecosystem. These methods have the advantage of being simple, inexpensive, easy to apply and basically require hydrological data collected at river gauging stations, which are often the only ones available for the region under study. The methods in this category use statistical hydrology tools such as the mean, median and permanence curve to provide recommendations for the minimum guaranteed flow (GONÇALVES, 2003; GALVÃO, 2008).

The use of hydrological models to support observations of the dynamics of the basins studied is extremely important for quantifying problems related to possible changes in the dynamics of the basins, as well as the availability of water to be used in various sectors, such as irrigation, thereby seeking the most efficient way of utilising water resources. In this sense, knowledge of the amount of water available in the Basin, as well as the demand for water resources, is essential in terms of providing subsidies for decision-making by public authorities.

In 2001, the Massachusetts Water Resources Commission carried out a study to define hydrological stress, which was defined as a basin or sub-basin in which the quantity of flow of a river has been significantly reduced, or its quality degraded (MWRC, 2001). In Tanzania, for example, the government has prioritised the use of water in its water policy. First for ecosystems and then for human uses. However, the calculation of the ecological flow is determined through the product of the social, ecological and economic values of water. In this country we see a disciplining of the competence to determine the ecological flow, which is defined by society but regulated by the state (ACREMAN et al., 2004).

The water policy used in Western Australia is based on the same principle as the model used in Tanzania. In that country, the Water and River Commission (Water Resources Agency) establishes that flows for ecosystems are prioritised. According to Postel and Richter (2003) water is first reserved to support ecosystems and only the remainder can be allocated to other uses. Postel and Richter (2003) point out that in Australia all the states have signed the Water Reform Framework agreement as a way of optimising the sustainable use and protection of ecosystems.

After the agreement was signed, consideration began to be given to balancing environmental, social and economic aspects when determining environmental flows. This led to the establishment of the Macro Water Sharing Plan in the country, which made it possible to manage the cumulative impact of extraction, facilitate the exchange of water titles and clarify the rights of the environment, users and water supply for the cities of New South Wales (NSW, 2006).

This makes it interesting to use a methodology that can use statistical methods to relate water demand and supply to physical parameters such as climate. This is why this study used an Australian methodology proposed by the Macro Water Sharing Plans (MWSP), which is capable of working with these parameters and, at the same time, adapting to the specific characteristics of the basin studied.

In Australia, the New South Wales Department of Natural Resources has been using this methodology since 2006 to determine the degree of hydrological stress in certain basins, taking as parameters to arrive at the final result the climatic vulnerability of the basin surveyed and the ratio of use of water resources in the area. According to the MWSP method used in Brazil by the National Water Resources Plan (PNRH, 2006), also applied by Galvão (2008) in the Ribeirão Piripau Basin, it is possible to estimate hydrological stress through the ratio between extraction demand and available flow (called Hydrological Stress - Eh).

This method will be useful as it will provide statistical data on the ratio of water resource use in irrigated areas, emphasising the relationship between water supply and demand in the Riacho do Pontal Basin. It will also identify the environmental vulnerability of the basin to climate change and estimate the degree of Hydrological Stress to which the water body is subjected. This survey will be important because this is a basin that will receive water from the transposition of the São Francisco River.

CHAPTER 3

DESCRIPTION OF THE STUDY AREA

3.1 Geographical Location

The UP13 Water Planning Unit, which corresponds to the Pontal stream catchment area, is located in the far west of the state of Pernambuco, between **08°19'00" and 09°13'24" south latitude and 4O°11'42" and 41°20'39" west longitude. The** Pontal stream has its source in the far west of the state of Pernambuco, between the borders of the states of Piauí and Bahia, in the municipality of Afrânio (BRITO et. al 2005).The Pontal stream basin has a drainage area of 6334 km at its mouth on the São Francisco river^2 , flowing into the left bank of the São Francisco river after travelling a distance of approximately 200 km, in a predominantly northwest-southeast direction (APAC, 2013), (Figure 1).

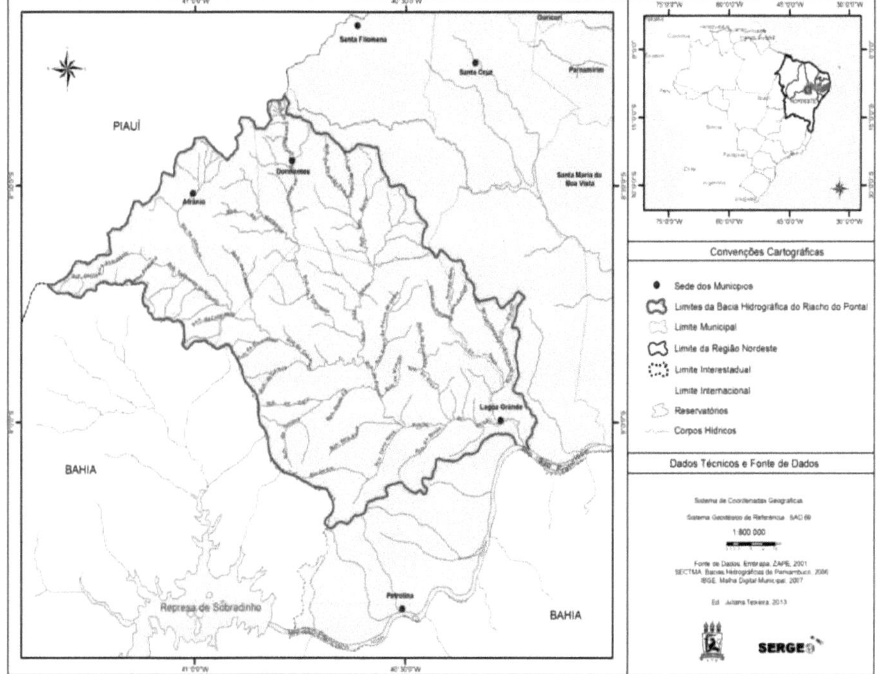

Figure 1 Location map of the Riacho do Pontal-PE watershed.

3.2 Hydrography

The Pontal Basin's main watercourses on the right bank are the Caieira, Sítio Novo, Terra Nova and Simão streams. On the left bank, the Caboclo, Caldeirão, Dormente, Baixo, Areial and Serra Branca streams stand out. The river's drainage area involves 4 municipalities, of which only the municipality of Afrânio is fully included in the basin (APAC, 2013).

The main tributary is the Dormente stream in the municipality of the same name, with a drainage area corresponding to 34% of the Pontal stream catchment area. A number of hydraulic projects (dams) have been built on these streams by the public authorities: Cruz de Salinas (4,021,375 m^3), Vira Beiju (11,800,000 m3) and Caititu (3,500,000 m3). The dams listed as part of the Pontal Creek Successive Dam System (public dams) are: Amargosa, Caldeirão II, Comprida, Gavião, Jatobá, Lajedo, Lagoa da Pedra II, Mandim, Poço da Serra and Poço do Canto (CODEVASF, 2007a).

3.3 Pontal Project

Initially, the choice of crops that would occupy the project area was based on the agronomic studies of the Pontal Project Feasibility Study. Considering market conditions, crop profitability, agro-industrial use, the area's vocation, crop adaptation to soil and climate conditions, food needs and socio-economic aspects. The following crops were selected by CODEVASF (2013):
- **main crops: pumpkin,** cotton, banana, sweet potato, beetroot, lemon, melon, corn, carrot, bean, guava, watermelon, pineapple, black mucuna, chilli and grape;
- **optional crops: acerola, asparagus, fig and mango.**

However, local observations and the agricultural occupation of the Senador Nilo Coelho and Bebedouro irrigated perimeters led to a reformulation of the species chosen. There has been a growing expansion of fruit growing in these irrigated perimeters, especially banana and mango crops. Therefore, it was established that 100 per cent of the irrigable area of each plot would be occupied by permanent species (fruit trees). The crops selected at this stage by CODEVASF (2013) were: - **permanent crops: banana, mango, coconut, guava and grape.**

3.4 Climate

The region's climate is classified as hot semi-arid (BSwh^5), with an average annual rainfall of 557.7 mm (BRASIL, 2004), with rainfall concentrated in four months (December to March). With an average potential evapotranspiration according to Hargreaves for the Petrolina station of around 2,090 mm per year, the average water deficit is 1,689 mm/year. It covers an area of 7,540 hectares, of which 4,029 ha are in studies or projects and 3,511 ha are in production, distributed in two areas separated by the Pontal stream: the South Area, with 3,511 ha, and the North Area, with 4,029 ha. The occupation includes 4,291 ha destined for 715 plots for small irrigators and 3,249 ha for 82 plots for medium-sized companies (CODEVASF, 2007a).

3.5 Vegetation cover

The vegetation in the area of direct influence of the project corresponds to a mosaic where shrubby caatinga predominates, interspersed with stretches of shrubby caatinga, the latter generally associated with some previous anthropogenic action. The municipalities of Petrolina and Lagoa Grande have the same physiognomy, with typical caatinga vegetation with a predominantly shrubby physiognomy with arboreal elements, which can be dense or open, with a shrub layer varying between 3 and 4 metres in height (CODEVASF, 2007b).

3.6 Soils

According to CODEVASF (2007b), the basin has soil classes such as Yellow Latosols, Red-Yellow Argillosols, Planosols, Cambissols, Quartz Sand Neosols and Litholic Neosols. Around 97,000 ha were studied for the implementation of irrigated perimeters in the area, of which around 54,000 ha were considered irrigable. These irrigable areas are alternative sites for the development of irrigated agriculture.

CHAPTER 4

MATERIAL AND METHODS

4.1 ESTIMATING HYDROLOGICAL STRESS

The study by Galvão (2008) was used as the main reference for this work. The step taken in this study corresponds to the calculation of hydrological stress for the Pontal Basin located in the Sertão region of Pernambuco. This indicator is the result of combining two parameters: Water Resources Use Ratio (Ru) and Vulnerability to Climate Change (Figure 2).

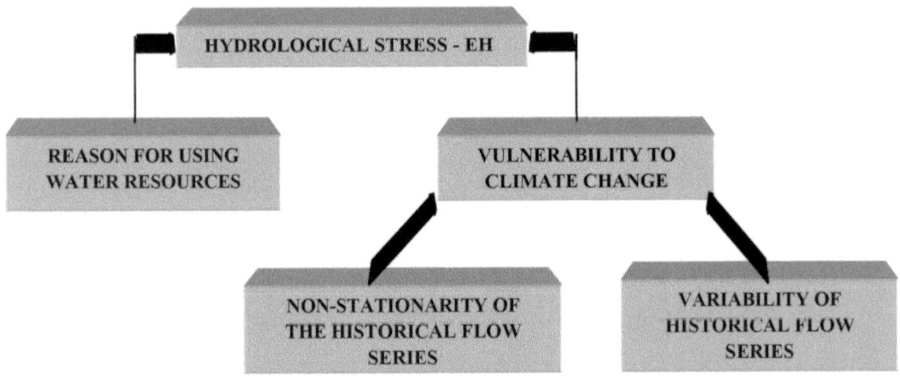

Figure 2: Parameters and sub-parameters that make up the Hydrological Stress - Eh indicator. Source: adapted from Galvão, 2008.

4.1.1 Water Resources Use Ratio - Stage 1

The Water Resources Use Ratio parameter relates the amount of water **available to the amount of water extracted from the watercourse, a similar criterion used** by the MWSP and the PNRH (Galvão, 2008). This parameter takes into account the demand for extraction, which in this study will be considered as the demand for water for irrigation (m³ /s) in the most critical month of the flow series studied, and the available flow as the long-term average of the flows (Equation 1).

(Equation 1)

$$Ru = {}^{Qd}\!/_{Qmed} \; x \; 100$$

In which:

Ru = Water Resources Use Ratio

Qd = Demand (m³ /s) in the most critical month

Qméd = Long-period average flow (m3/s)

Determining the average flow is important in a basin because it represents the maximum water availability. It is the highest flow that can be regularised, allowing the assessment of the upper limits of the use of water from a spring for different purposes (TUCCI, 2002).

Once the use ratio had been calculated, its value was examined: if the percentage of use of water resources is less than 20 per cent, its use ratio is considered low, if the value is in the range between 20 and 50 per cent, the use ratio is medium and if it is above 50 per cent, it is high, as shown below (Table 1).

Table 1 - Parameter Use Ratio of Water Resources in the Pontal Hydrographic Basin. Source: GALVÃO, 2008.

REASON FOR USE	% use of Ru
Bass	RU < 20%
Medium	**20 ≤RU≤50%**
High	RU > 50%

The second parameter for measuring hydrological stress is the Vulnerability to Climate Change parameter, which characterises the vulnerability of the river basin to climate change, classifying it into three levels: low (B), medium (M) and high (A). It is calculated using two sub-parameters that check the coefficient of variation of the historical series of flows (variability), and the degree of stationarity of the historical series of flows. These sub-parameters are climate variability and the non-stationarity of the historical flow series.

4.1.2 Climate Variability - Stage 2

In this work, variability is estimated by means of the coefficient of variation of the flows over the period studied. This is calculated using the standard deviation (SD), which is given by the formula below (Equation 2):

Equation 2

$$\sigma = \frac{R}{d2}$$

Where R is the average amplitude of the process and d2 is the factor related to the size of the subgroups. Once the historical series of average annual flows for the **studied watercourse had been obtained, the coefficient of variation (CV)** *was* **calculated using** Equation 3, where the standard deviation is divided by the average flow of the basin in the years studied and multiplied by 100.

Equation 3

$$CV = \frac{standard\ deviation}{average} \cdot 100$$

In statistics, the *coefficient of variation is* a measure of the dispersion of a sample in relation to its mean. It is useful because it extends the analysis of the standard deviation, providing a relative measure that is independent of the quantity with which the sample data is measured. It is used, among other things, to compare the relative dispersion of two samples, even if their means or magnitudes are different, or to measure the risk of an investment: the lower the coefficient of variation of its returns, the lower its risk. The coefficient of variation is dimensionless, since the standard deviation of the sample has the same dimensions as the mean. The value found must be compared to those in Table 2, thus defining the level of variability for the management unit being assessed. If the value found is less than 15 per cent, it is low, if it is between 15 and 30 per cent, it is medium and if it is greater than 30 per cent, it is high.

Table 2 - Variation of the sub-parameter Variability of the Historical Flow Series.

VARIABILITY	CV
Bass	< 15%
Medium	15% ≤ CV ≤ 30%
High	> 30%

4.1.3 Non-stationarity of the historical flow series - Stage 3

The second sub-parameter to be assessed is the stationarity of the flow series, which can be estimated using the Salas test (1993). According to this test, a flow series is considered stationary if it is free of trends, variations or periodicities. In other words, a stationary series means that the statistical parameters of the series, such as mean and variance, remain constant over time. Thus, it is assumed that the sequence of hydrological data, whether flow or precipitation, is statistically stationary in the sense that the values of the sequence fluctuate randomly around a mean value that remains constant over time, and that the dispersion of the data around the mean also remains constant. Otherwise, according to CLARKE (2003), the series is considered non-stationary. In this sense, the non-stationarity test for flow series is given by Equation 4 derived from Salas (1993):

Equation 4

$$tc = \frac{r\sqrt{N-2}}{\sqrt{1-r^2}} > t1 - \frac{\alpha}{2}, v$$

Where: tc (stationarity index) = value of t

r = correlation coefficient between q(i) (parameter) and number of years studied (i) (time). The square of Pearson's correlation coefficient is called the coefficient of determination or simply r^2 and is a measure of the proportion of the variability of one variable that is explained by the variability of another. It is unusual to have a perfect correlation ($r^2 = 1$) in practice, because there are many factors that determine the relationships between variables in real life.

N = number of years in the series

v = N- 2 degrees of freedom

t1-α/2,v = *Student*'s t at 99% confidence (tabulated value of Student's distribution for one degree of freedom n = N-2). The value obtained in the test is evaluated according to Table 3.

Table 3. Variation of the Non-Stationarity sub-parameter of the Historical Flow Series.

NON-STATIONARITY	
LOW	tc < 0.9 t
MEDIUM	0.9t≤tc≤ 1,lt
HIGH	tc > 1.1 t

Where a tc less than 0.9t (0.9 x Student's t) represents low non-stationarity, i.e. either the series is stationary or its degree of non-stationarity is small. If the tc is in a range between 0.9t and 1.1t, it is considered to be of medium non-stationarity and if it is above 1.1t (1.1 x Student's t), the historical flow series has a high degree of non-stationarity. It is important to emphasise that when tc is less than t, the series shows no significant upward or downward trend and is therefore considered stationary (CHAVES et. al., 1997).

4.1.4 Climate Vulnerability - Stage 4

After finding the values of these two sub-parameters, they will be combined to create the parameter of vulnerability to climate change. These values (Table 4) were created using the three levels and their corresponding scores for each indicator (Low = 1, Medium = 2 and High = 3), parameter and sub-parameter. Each element in the table is the product of the scores corresponding to the levels in the row and column (GALVÃO, 2008).

Table 4. Combination of the Variability and Non-Stationarity sub-parameters of the Historical Flow Series.

	NON-STATIONARITY		
	LOWER MIDDLE		HIGH
LOW	12		3

| VARIABILITY | MEDIUM | 24 | | 6 |
| | HIGH | 3 | 6 | 9 |

The integration of the previous Variability and Non-stationarity sub-parameters (Table 4) enables the final calculation of the Vulnerability to Climate Change parameter (Table 5).

Table 5 - Variation in the Level and Score for the Vulnerability to Climate Change parameter.

VULNERABILITY TO CLIMATE CHANGE	SCORE
LOW	1 -2
MEDIUM	3-4
HIGH	6-9

By combining the parameters Water Resources Use Ratio and Vulnerability to Climate Change (Table 6), it is possible to obtain the Hydrological Stress indicator.

Table 6 - Combination of Water Resources Use Ratio and Vulnerability to Climate Change parameters

| | | RATIO OF USE OF WATER RESOURCES - Ru | | |
		LOW	MEDIUM	HIGH
	LOW	1	2	3
Vulnerability to Climate Change	MEDIUM	2	4	6
	HIGH	3	6	9

Hydrological stress - Stage 5

Based on the work of Galvão (2008), the higher the Vulnerability to Climate Change and the Ratio of Use of Water Resources, the higher the Hydrological Stress - Eh and the more vulnerable the river basin is with regard to water supply and demand and susceptibility to severe climatic phenomena.From the results obtained in Table 6, the relationship with the corresponding Score is carried out, so it is possible to evaluate the final value for the Hydrological Stress - Eh indicator (Table 7).

Table 7. Variation in Level, Score and Final Value for the Hydrological Stress - Eh indicator.

LEVEL Eh	SCORE	VALUE
LOW	1 -2	1
MEDIUM	3-4	2
HIGH	6-9	3

Based on this methodology, and after working with the fluviometric data from the BEBEDOURO station, located in the municipality of Petrolina, in the Pontal basin, in the hinterland of the state of Pernambuco, it was possible to calculate the **hydrological** stress that **the watercourse in this study area has been experiencing.**

CHAPTER 5

RESULTS AND DISCUSSION

5.1 Application of the MWSP methodology

By calculating the ratio of use of water resources as shown in equation 1, in which the Ratio of Use of Water Resources is given by the ratio between the demand in the most critical month, in terms of the demand for water for irrigation and the average flow over the long period, in this sense, It should be noted that there is no information available on the monthly demand in the most critical month, but by calculating the annual demand, which according to PERH/PE (1998) is *202.454*106m³ /year, the* demand was divided by 12 months = 16.87x106m3/year = 5.35x10 x10^{-76} m3/s = 0.535 m3/s, so there is an average monthly consumption of 0.535 m3/s.

Qméd = Long-period average flow (m3/s) = 1.04098m3/s Ru = Qd/Qmed x 100= 51.394%.

In other words, even though this is not the most critical month in terms of water consumption, the average exceeds the use ratio considered to be high (above 50% of consumption for irrigation alone) (Table 10).

Table 10. Results for the Water Resources Use Ratio parameter in the Pontal Hydrographic Basin.

REASON FOR USE	% use of Ru
Bass	RU < 20%
Medium	**20 ≤RU≤50%**
High	RU > 50%

Source: GALVÃO, 2008.

The use of the water resources use ratio parameter shows that it is impossible for the Basin to supply enough water to meet the needs of its users, focussing mainly on the use of water for irrigated crops.

Using the flow data inferred by SECTMA for the years 1935 to 1985 (ANNEX I), the average annual flow was calculated, resulting in an average of 1.04098. The variability of the historical flow series was then estimated (Figure 3).

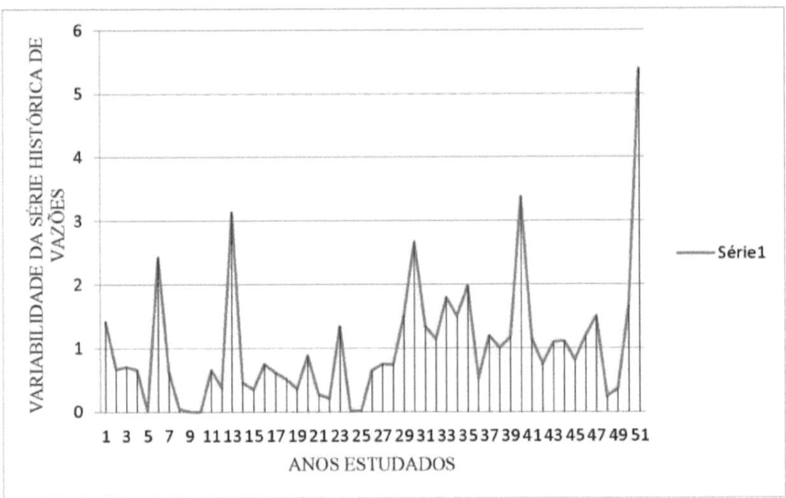

Figure 3 - Variability of the historical flow series (1935-1985).
Source: adapted from SECTMA.

After calculating the average of the years sampled, the standard deviation (SD) and coefficient of variation (CV) were calculated, where the standard deviation (SD) is divided by the average flow of the years under study and multiplied by 100, resulting in a CV of around 94.82%. Below is the result for the sub-parameter on the variability of the historical series of flows in the Pontal-PE Hydrographic Basin (Table 11).

Table 11 - Result (hatched) for the sub-parameter Variability of the Historical Series of Flows in the Pontal-PE Hydrographic Basin.

VARIABILITY	CV
Bass	< 15%
Medium	15% ≤ CV ≤30%
High	> 30%

Variability is therefore considered high in the Pontal Basin because the CV is above 30 per cent. A result of high variability was also found by Silva and Galvíncio (2011) in the Ipojuca-PE Basin in Agreste Pernambucano, showing that there are similarities in the variability of the historical series of flows in these basins that will receive water from the transposition of the São Francisco River. In the study by Galvão (2009), the coefficient of variation of the historical series of flows was 34.88%, considered high, but in comparison with his study, the result of this work indicates an even higher coefficient of variation, which resulted in a high variability of the series.

Subsequently, the non-stationarity test (to find out if the series is stationary or not) was calculated using the formula in equation 4. It should be noted that the data obtained in this work has a sampling period of 50 years, as the above test is based on Student's t-test at 99% confidence (tabulated value of Student's distribution for one degree of freedom n = N-2), the degree of freedom was 0.013, meaning that the probability of -oo < t < 0.013 is 99%. The r in the formula, which is the tendency to increase or decrease, is calculated using the flow data, resulting in a value of 0.3439. The realised tc is shown in Table 12:

Table 12 - Calculated values for the SALAS method (1993).

AVERAGE	1,041
D.P.	0,987
C.V.	94,82%
t	0,013
r	0,3439
Tc	2,53

Thus, according to the values of tc in relation to 0.9t and 1.1t, tc has a value greater than 0.9t and 1.1t, and it can be concluded that the historical series shows a low degree of non-stationarity in the Basin (Table 13).

Table 13. Degree of non-stationarity.

tc	*t(Student)*	0,9t	1,1t	Non-stationarity
2,53	0,013	0,0117	0,0143	High

From the calculation in table 13, the corresponding value must be hatched in the non-stationarity table below (table 14).

Table 14. Result (hatched) for the Non-Stationarity sub-parameter of the Historical Qmed Flow Series (1935-1985) in the Pontal-PE Hydrographic Basin.

NON-STATIONARITY	
LOW	tc < 0.9 t
MEDIUM	0.9t≤tc≤ 1,1t
HIGH	tc > 1.1 t

In the case of the Riacho do Pontal Basin, as discussed above, non-stationarity was considered high in the periods from 1935 to 1985, so the value of tc (which was above 1.1t) is considered high. The same result of non-stationarity was found in the work by Silva and Galvíncio (2011) in the Ipojuca-PE Basin, which may indicate a trend in the historical series in the region's basins.

This result differs from that found by Galvão in the Ribeirão Piripau Basin (DF/GO), where non-stationarity was considered low, meaning that there the historical series showed little or no trend within a characteristic variability in the region, i.e. there were no significant changes in the statistics of the historical flow series.

From the two sub-parameters (variability and non-stationarity) we can cross-reference (or integrate) them and calculate the Basin's Vulnerability to Climate Change parameter (Table 15).

Table 15. Combination of the Variability and Non-Stationarity sub-parameters of the Historical Flow Series.

		NON-STATIONARITY		
		LOW	MEDIUM	HIGH
	LOW	1	2	3
VARIABILITY	MEDIUM	2	4	6
	HIGH	3	6	9

The higher the Variability and Non-Stationarity sub-parameters, the greater the vulnerability of ecosystems and the society that makes use of these resources to extreme events (GALVÃO, 2008). Thus, climate variability was considered high while non-stationarity was considered low, being represented by the product of the two (3 x 3 = 9) as shown in Table 16:

Table 16. Results (hatched) for the vulnerability to climate change parameter for the period 1935-1985.

VULNERABILITY TO CLIMATE CHANGE	SCORE
LOW	1 -2
MEDIUM	3 -4
HIGH	6-9

Therefore, combining variability with non-stationarity resulted in a high score (9) for vulnerability to climate change (6-9).

By integrating the parameters Use Ratio and Vulnerability to climate change (Table 17), the Basin's hydrological stress (Eh) can be assessed:

Table 17. Combination of Water Resources Use Ratio and Vulnerability to Climate Change parameters. Source: adapted from Galvão (2008).

		RATIO OF USE OF WATER RESOURCES - Ru		
		LOW	MEDIUM	HIGH
	LOW	1	2	3
Vulnerability to Climate Change	MEDIUM	2	4	6
	HIGH	3	6	9

As Ru was high (3) and vulnerability was high (3), the product of these two is the number 9, so this number is compared with the level of Hydrological Stress, as shown in the table below (Table 18):

Table 18. Result (hatched) of the Level Variation, Score and Final Value for the Hydrological Stress - Eh indicator.

LEVEL Eh	SCORE	VALUE
LOW	1 -2	1
MEDIUM	3 -4	2
HIGH	6-9	3

Source: adapted from Galvão (2008).

Therefore, by combining variability with non-stationarity, a high score (9) was obtained for vulnerability to climate change, and by combining the Use Ratio of water resources with vulnerability, a high overall score (9) was obtained for the hydrological stress indicator (Eh = 3).

It is worth noting that hydrological stress in the Pontal-PE Basin was considered high, as it was in the Ipojuca-PE Basin in a study carried out by Silva and Galvíncio (2011). Thus, the results of hydrological stress obtained by holistic methods tend to indicate high hydrological stress in the basins of the Brazilian semi-arid region when compared to other regions of the country. As we have seen throughout this paper, the Pontal-PE basin has high variability in terms of inter-annual and intra-annual flows, as can be seen in Annex I of this paper. Given its intermittency, in some months of the year this flow reaches zero, which jeopardises the supply both for agriculture, which is the focus of this work, and for other advisory uses.

If we compare these results with those obtained by Galvão (2008) in the Piripau River Basin (DF/GO), we can see that this study used practically the same methodology as the author. A similar study was also carried out by Silva and Galvíncio (2011) in the Ipojuca River Basin in the state of Pernambuco, which is also one of the basins that will receive water from the transposition of the São Francisco River. In this study, it was found that the Ipojuca River Basin in the state of Pernambuco was highly vulnerable to climate change, similar to what was observed in the Pontal Creek Basin in the state of Pernambuco up until 1985. In Galvão's work (2008), which is one of the few to use the methodology in the country, the author arrives at a result of average hydrological stress in the Piripau River Basin (DF/GO). The author explains that the average result of hydrological stress reflects a possible climate change scenario, changes in land use in the Basin and increased water extraction could aggravate this situation.

For this reason, integrating the Riacho do Pontal-PE Basin with the São Francisco River is the most efficient alternative to the problem of intermittency in the Basin. This study therefore presents complementary results to those presented in the EIA/RIMA for the Pontal-PE Basin, as it can be seen that the EIA/RIMA uses flow data from the Basin, but without the appropriate methodology to obtain other results that quantify the real need for water from agricultural crops and without measuring susceptibility to changes in climate patterns.

In this sense, the work deserves attention because it concerns a Basin that will receive water from the transposition of the São Francisco River, thus incorporating, in addition to the natural climatic susceptibility of the Basin, negative impacts resulting from this transposition. Given this reality, the transposition of the São Francisco River appears to be a viable and beneficial alternative for agriculture in the study area, with an increase in available water and a reduction in losses due to reservoirs, with a potential increase in agricultural production and, consequently, economic dynamism in the area, thus reducing the problems caused by drought, such as food shortages, low productivity in the countryside and rural unemployment.

The natural dynamics of the Pontal Creek Basin in the state of Pernambuco will be affected by the works to transpose the São Francisco River and by the flow of water coming from the implementation of this project. This will create new opportunities for land use and positive and/or negative impacts on nature in areas that did not receive water or received it at insufficient levels. Therefore, these studies provide support both for the methodological part and for the actual results themselves, because in light of the above, it is clear that this Basin has shown a certain fragility with regard to its own flow dynamics and the maintenance of the natural characteristics of the caatinga vegetation, and climate change tends to accentuate the effects on the population and economy of the surrounding areas. In order to prevent this increase in water supply from causing problems linked to soil salinity and water wastage, the aim of this study was to estimate the optimum water table required for each agricultural crop to be grown in the Basin.

CHAPTER 6

CONCLUSIONS

> It was concluded, using the water resource use ratio parameter, that focusing mainly on the use of water for irrigated crops, it is impossible for the Basin to provide enough water to meet its users, so that integrating the Riacho do Pontal-PE Hydrographic Basin with the São Francisco River Basin is the most viable alternative for the region's reality.

> The statistical evaluation of the flow data from the Pontal Creek Basin in Pernambuco showed that the area studied has a high degree of climate variability. The results on climate variability are noteworthy because data from the literature shows that there are implications of climate variability and change for water resources: greater variance implies the need for greater resilience in water infrastructure (reservoirs, canals, pumping stations, etc.), i.e. more complex and costly systems; and it signals that the water supply systems of small communities and metropolitan regions are operating, or are about to operate, at the limit of their capacity. High climate variability is therefore associated with the occurrence of more severe extreme events than in regions with lower variability.

> The statistical evaluation of the flow data yielded a result of high non-stationarity, thus denoting a high tendency for the flow to increase or decrease during the series of years studied. The literature shows that non-stationary hydrological series can signal changes in land use of various kinds such as: deforestation, use of different agricultural practices, withdrawal of water for advisory uses (mainly irrigation) or even demonstrate climate change due to the greenhouse effect.

> By combining the data on non-stationarity and climate variability, it was also found that the Basin is highly vulnerable to climate change. This result shows that there is a need to try to find ways of mitigating the possible environmental impacts of this process in the Pontal-PE Basin.

REFERENCES

ACREMAN, M.; KING, J.; HIRJI, R.; SARUNDAY, W. and MUTAYOBA,W. **Capacity building to undertaking environmental flow assessments in Tanzania.** 2004, 11p. Available:http://www.iwmi.cgiar.org/research_impacts/Research_Themes/BasinWater Management/RIPARWIN/PDFs/Mike%20Acreman%20EF%20cap%20build%20Tanza nia%20paper.pdf.

ARAGÃO, T. G. **Transposition of water from the São Francisco River to the Paraíba River basin: an assessment of synergy and water sustainability using the Acquanet flow network model.** Dissertation presented to the postgraduate programme in Civil and Environmental Engineering. Campina Grande-

PB. April 2008.

BRAZIL. MINISTRY OF AGRICULTURE AND SUPPLY. **Programme to support and develop irrigated fruit growing in the northeast (PADFIN).** Brasília: SPI, 1997. 148 p. (Basic Document).

BRAZIL. Pontal Project - North Area - Executive Project - **Pedological Services for Drainage Purposes of Patches 20 and 23** Volume 4.3. 1998.

BRAZIL. Pontal Sul Project - Executive Project - **Pedological and Drainage Service Report** - Vol. 4.3, 1998.

BRAZIL. MINISTRY OF NATIONAL INTEGRATION. **Inventory of Projects of the São Francisco Valley Development Company** - CODEVASF - Third Edition.Brasília. 1999. Page 90.

BRAZIL. MINISTRY OF NATIONAL INTEGRATION; CODEVASF; NORONHA - TAMS CONSORTIUM. **Pontal Project - North Area located in the municipality of Petrolina, state of Pernambuco. Executive Project. Final Project Report.** V.5.2 (Specific Reports - Environmental Studies. Tome I - EIA and Tome II - RIMA), Dec./2000.

BRAZIL. MINISTRY OF NATIONAL INTEGRATION; SÃO FRANCISCO AND PARNAÍBA VALLEYS DEVELOPMENT COMPANY - CODEVASF; NORONHA CONSORTIUM - TAMS. **Pontal Project - North Area located in the municipality of Petrolina, state of Pernambuco.** Executive Project. Final Project Report. Vol. 5.2 (Specific Reports - Environmental Studies. Volume I - EIA and Volume II - RIMA), Dec. 2000.

BRAZIL. MINISTRY OF NATIONAL INTEGRATION - Water Infrastructure Secretariat - **Water Infrastructure Studies and Works.** Brasília, June 2001. Page 234.

BRAZIL. MINISTRY OF NATIONAL INTEGRATION / CODEVASF - **Mais Irrigação is launched by President Dilma with forecast investment of R$10bn in 538,000 hectares and 16 states.** In: http://www.codevasf.gov.br/noticias/2007/mais-   irrigacao-e-lancado-com-previsao-de-investimento-de-r-10-bi-em-538-mil-hectares-and-16-stados. Brazil, 2007a.

BRAZIL. MINISTRY OF NATIONAL INTEGRATION / CODEVASF - **Environmental Impact Study - Pontal Norte Project.** Projetec. Volume 1, preliminary studies, Brazil, 2007b.

BRAZIL. MINISTRY OF NATIONAL INTEGRATION / CODEVASF - **Environmental Impact Study - Pontal Norte Project.** Projetec. Volume 2, preliminary studies, Brazil, 2007c.

BRAZIL. MINISTRY OF NATIONAL INTEGRATION. **Irrigation in Brazil Situation and Guidelines.** BRASILIA, 2008.

BRAZIL. BRAZILIAN AGRICULTURAL RESEARCH CORPORATION - EMBRAPA. Organiser: Ana Alexandrina Gama da Silva. SILVA, A. A. G. ; AMORIM, J. R .A.; FACCIOLI, G. G. ; SOUSA, I. F.; MILET, W. B. ; ROCHA, J. C. S. ; ROCHA, A. F.; **Teaching material for the practical course on "Agrometeorology applied to optimising the use of water in irrigation".** ISSN 1678-1953. December 2009.

BRAZIL. MINISTRY OF SOCIAL DEVELOPMENT AND FIGHT AGAINST HUNGER.

Opportunities and possibilities for the development of family and urban agriculture in the local context. Org. Ana Ruth Silva. Petrolina, September 2010.

BRAZIL. MINISTÉRIO DA INTEGRAÇÃO NACIONAL / CODEVASF - **Irrigation in the World - Precedents from Other Countries -** (Source: FAO), 2013.

BRAZIL. MINISTRY OF NATIONAL INTEGRATION / CODEVASF - **The Concession Law & Public-Private Partnership - PPP in the Irrigation Sector,** 2013.
FRIZZONE, J. A. **Optimisation of water use in irrigated agriculture:** perspectives and challenges. Engenharia Rural, v.15, único, p.37-56, 2004.

GALVÃO, D. M. O. 2008. **Subsidies for determining environmental flows in unregulated watercourses:** the case of Ribeirão Piripau (DF/GO). Master's dissertation, University of Brasilia, Department of Forestry Engineering, Publication PPGEFL. DM - 096/08, Brasília, DF.

GONÇALVES, M. V. C. 2003. **Methodology for determining minimum guaranteed flows in watercourses.** Master's dissertation, University of Brasília, Department of Civil and Environmental Engineering, Publication MTARH-DM - 061/03, Brasília, DF, 129p.

MORAES, A. C. R. **Geografia:** pequena história crítica. São Paulo: Hucitec, 1987.

NEW SOUTH WALES. **Macro Water Sharing Plans:** the approach for unregulated rivers. Australia: Department of Natural Resources - NSW. 56p. 2006.

PERNAMBUCO. Pernambuco Food Supply Centre - CEASA. **"More Irrigation"** **for Pernambuco .** 2012. In: http://www.ceasape.org.br/verNoticia.php?id=1570.

PERNAMBUCO. Pernambuco State Water Resources Plan - PERH/PE. **The state water resources policy.** Government of Pernambuco, volume 8, 1998.

PINEDA, L. A. C. **Observational and Hydrological Modelling Study of a Micro-Basin in Undisturbed Forest in Central Amazonia.** 241f. Thesis (Doctorate in Meteorology). INPE. São José dos Campos, 2008.

POSTEL, S.; RICHTER, B. **Rivers for life:** Managing water for people and nature. USA: Island Press. 2003, 253p.

REBOUÇAS, A. C. Freshwater in the World and in Brazil. In: REBOUÇAS, A. BRAGA, B. TUNDISI, J.G. (Org). Fresh **Waters in Brazil:** Ecological Capital, Use and Conservation. São Paulo: Escrituras, 2nd Ed. Revised and Expanded, 2002.

REBOUÇAS, A. C. **Water in Brazil:** abundance, waste and scarcity. BAHIA ANÁLISE & DADOS Salvador, v. 13, n. ESPECIAL, p. 341-345, 2003.

REBOUÇAS, A. C. **Intelligent Use of Water.** São Paulo: Escrituras, 2004.

SANTOS, M. **Por uma Geografia nova:** da crítica da geografia a uma geografia crítica. São Paulo, HUCITEC, 2002.

SETTI, A. A.; LIMA, J. E. F. W.; CHAVES, A. G. de M.; PEREIRA, I. de C. (2001). **Introdução ao gerenciamento de recursos hídricos.** 2. ed. - Brasília: Agência Nacional de Energia Elétrica, Superintendência de estudos e informações hidrológicas.

SILVA, E. R. A. C.; MIRANDA, R. de Q.; FERREIRA, P. dos S.; GOMES, V. P.; GALVÍNCIO, J. D. Estimação do Estresse Hidrológico na Bacia Hidrográfica do Riacho do Pontal-PE / Hydrological stress estimate in Pontal watershed-PE. **Caderno de Geografia,** v.26, n.47, 2016. DOI: https://doi.org/10.5752Zp.2318-2962.2016v26n47p844

SILVA, E. R. A. C.; GALVÍNCIO, J. D.; NASCIMENTO, K. R. P.; SANTANA, S. H. C. de; SOUZA, W. M. de; COSTA, V. S. de O. Analysis of the temporal trend of inter-annual and intra-annual rainfall in the semi-arid region of Pernambuco. **Brazilian Journal of Climatology.** Vol. 22, 76-98. Jan/Jun, 2018. http://dx.doi.org/10.5380/abclima.v22i0.53956

Printed by Books on Demand GmbH, Norderstedt / Germany